DO YOU READ ME?

VINTAGE COMMUNICATION TOYS

Schiffer Publishing Ltd
4880 Lower Valley Road • Atglen, PA 19310

LESLIE SINGER

Other Schiffer Books on Related Subjects:

Compendium of Automatic Morse Code, by M. Edwin Goss, 978-0-7643-5131-0
Retro Radio: Six Decades of Design 1920s–1970s, by Mike Tauber, Foreword by Ron Simon, 978-0-7643-4679-8
Deco Radio: The Most Beautiful Radios Ever Made, by Peter Sheridan, 978-0-7643-4605-7

Library of Congress Control Number: 2018934127

Designed by Patti Woods Rye
Cover design by Patti Woods Rye & Matthew Goodman
Type set in Streetwear/Futura Std

ISBN: 978-0-7643-5578-3
Printed in China

Published by Schiffer Publishing, Ltd.
4880 Lower Valley Road
Atglen, PA 19310
Phone: (610) 593-1777; Fax: (610) 593-2002
E-mail: Info@schifferbooks.com
Web: www.schifferbooks.com

To my children, Hannah, Noah, and Kate.

And to their children, Cici, James, Ava, and Collins—who just assume a phone can take pictures, be your game partner, and guide you to any place in the world.

Space Phone
Hale-Nass Corp., USA

Electronic Walkie Talkie
Mfg. unknown

Contents

Illustration by Chad Maupin

About the collection

I became fascinated with toy walkie-talkies and other communication toys when I was around ten years old. This was in the early-'50s, the same time I came to love ray guns. (This is a good place to mention that I wrote *ZAP! Ray Gun Classics*, Chronicle Books, 1992.) Both shared two magical elements—action and the future!

Graphics representing the future always thrilled me. These came in the form of packaging design, comics, trading cards, and other imaginative visions. Unlike some collectors, I'm not in the hunt for "mint" items. Even though they demand a higher price in the collector market, I actually prefer toys that proudly wear their battle scars, proof that they have lived their life as intended.

I do not represent this collection as being complete. Some toys are not included, either because I didn't have enough interest to own them, they were too similar to others, or I just never came across them. I also have not included very many of the later ones, which would actually broadcast signals over the air.

I've tried to group these by decade and to list the manufacturer when I know it, but I may be wrong in some cases. Obviously, this is not meant to be a scholarly work, far from it. It's meant just for fun, with an emphasis on the design of these wonderful creations.

And finally, for those too young to know or remember, the question "Do you read me?" was used by actual walkie-talkie or other radio communication operators when sending messages. It meant something like "Can you hear me now?"

Tom Corbett jigsaw puzzle
Saalfield Publishing Co.

Acknowledgments

Huge thanks to Willie Allen for his beautiful photography and Larry Pennington for the use of his North Little Rock studio and equipment. Endless appreciation to Patti Woods Rye for her exquisite page design and longtime collaboration. To Rick Armellini for his really cool paper engineering of the build-it-yourself Singer Communicator found in this book, and for his gifts, friendship, and inspirations over the years. Special thanks to Steve Joy for the gift of his fabulous handmade Farnsworth Communicator and for his patience. To Dr. Renie Bressinck for his Morse code acumen. To Melinda Woods for sharing some of her treasures. To Kirby Shofner for his assistance. And to Eric Wrobbel and Duane Dimock for photographing and sharing items from their collections.

Box detail from the Japanese Communication Set by Modern Toys

Tuning in

These days, when almost everyone on our planet seems to exist via their "smart" phone, it's nice to recall when there were similar devices that were, if not exactly "dumb," certainly not as evolved.

Among these were radios, telephones, and perhaps the most exotic—walkie-talkies. Its name said it all. You could talk and listen on them, but amazingly, you could walk around while you were communicating! Some were so small, you could wear them on your wrist. Astounding!

Unlike today's devices, they were rarely if ever used for idle chit-chat. Walkie-talkies were for serious business only: emergencies, rescue efforts, calling in air strikes, that kind of thing.

The little brother of the walkie-talkie was, of course, the toy walkie-talkie. With the exception of the iconic tin cans connected by a piece of string, the earliest ones I have found were made of paper, often with a thin cord of some sort connecting them—the transmission line to carry some feeble sound vibrations from one "phone" to the other. But among my favorites were the ones that didn't really work at all. No string or wires, no batteries, and nothing but the thread of imagination to save your life, warn your base of approaching danger, or save your planet from horrific aliens.

These could not be used to see what the weather was going to be, find a recipe, or watch people having sex. These were serious-looking devices for serious play, all the more serious if they had lightning bolts printed or painted on them. These meant their signal was stronger, their need more urgent.

Walki-Talki
DCMT Ltd./Lone Star Products, England

Higher frequencies

Just prior to World War II, and certainly after it, toy walkie-talkies exploded in popularity. Buck Rogers, that phenomenal predictor of our future, used all manner of them (along with his amazing ray guns), and his communicators are among my favorites. Many of these, especially the later ones, were created using that new miracle material, plastic!

Plastic took the design of the toy walkie-talkie, and just about everything else, into an entirely new dimension of imagination. Along with the new looks came the use of batteries and wire. But this capability was beside the point. Actually, batteries and wire barely improved the transmission quality over string. What's more, the advertising and packaging often never even mentioned the need for wire at all. How many kids' hearts sank when they opened the box of their new Dick Tracy Wrist Radios promising "fantastic electronic communication," only to find they must be tethered together by fifty feet of wire?

My personal, all-time favorite is the Remco Radio Station (pg. 13). It was not, technically speaking, a walkie-talkie because it was a "table-top" unit. But what a beauty of design and function. It had everything! In addition to its two microphones (and several yards of wire), there was a rotating searchlight controlled by a compass dial. Another dial gave you the Morse code, which could be tapped out by a code key. The code could be sent as telegraph clicks, buzzes, or searchlight beams depending on which of the switchboard ports you plugged into. There was even a tiny note pad to record the message as it arrived, should your friend be fortunate enough to have a Remco Radio Station of his or her own.

Walkie Talkie Phone
A Russell Product, USA

This was the case with my childhood best pal, Jeff Adler, who lived across Curley Street from me in Long Beach, New York, a fellow space cadet who flew and fought with me through the galaxy. We would each set up our Remco Radio Stations in our facing home windows and signal each other. Make that, *try* to signal each other. After much excited waiting for dark so we could tap out a few flashes from our search lights, one of us would have to call the other on the phone. "Was that two dashes and a dot, or three dashes, or three dots, or . . ." This "follow-up" messaging went on after virtually every series of light bursts. And I still remember his phone number more than sixty years later. Ah, the power of communication!

Today, I don't even know my kids' phone numbers. My phone knows them for me, and their email, and their birthdays, and my appointments, and virtually everything there is to know in the world. So here's to the past and to imagination. When a block of wood allowed you to talk to someone in a whole other solar system.

Leslie Singer
January 2017

Remco Industries deserves a special "Loud and Clear" award for their imaginative and beautifully designed toys of all kinds, but especially for their communication toys. Remco was founded in the 1940s by two cousins, Ike Heller and Saul Robbins. Armand Daddis soon joined the two as the company gradually moved from simple "walkie-talkies" to a variety of sophisticated remote-control toys. The name "Remco" comes from the two words "Remote Control." Originally located in Newark, New Jersey, the company later moved to nearby Harrison, New Jersey. Remco filed for bankruptcy in 1971 and was acquired by Azrak-Hamway International, Inc., in 1974. In 1997 Jakks Pacific acquired Remco from Azrak-Hamway.

Source: Wikipedia

An almost identical toy, the Dan Dare Space Control Radio Station, was manufactured in England by Randalls under their Merit brand, probably under license from Remco.

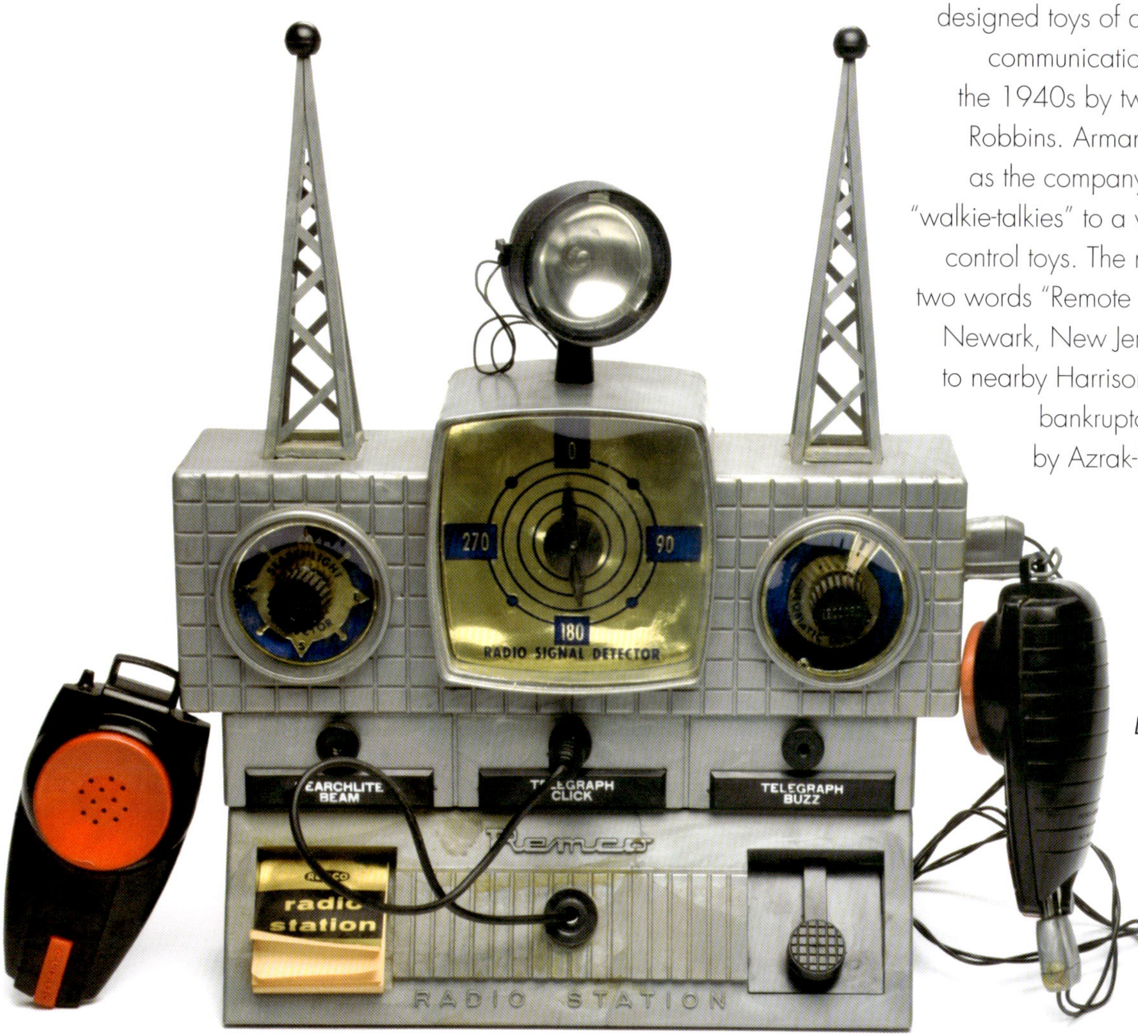

The beautiful Remco Radio Station
Remco Industries, USA

Biekens Biscuits premium card
England

Not only did this card predict the invention of the mobile phone—it predicted that people would ignore each other while using it during their social visits.

1930s

Buck Rogers cartoon illustration (copy)
Artist: Dick Calkins
John F. Dille, Co., 1931

Radio-Phone and box
Steel Stamping, Co. USA

Biekens Biscuits premium card
England

Two-Way Radio
Mfr. unknown, Germany

Radio Tinker
The Toy Tinkers, USA

Little Wonder microphone and box
Wonder Specialties, USA

Flash Gordon 2 Way Telephone
Louis Marx & Co., USA

Phon-o-Mike
Goody, USA

A piece of string, some cardboard and imagination. Do you read me?

Pillsbury's Farina Tel-A-Phone
Mfr. unknown

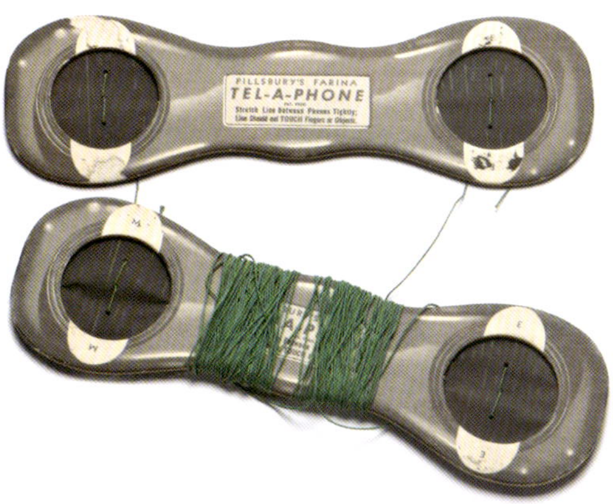

Tums walkie talkie
Lewis Howe Corp, USA

Micro-Fone
Mfr. unknown

Radio Mike
Mfr. unknown

Coco-Wheat Mike-O-Fone
Mfr. unknown

1940s

Telephones
Mfr. unknown

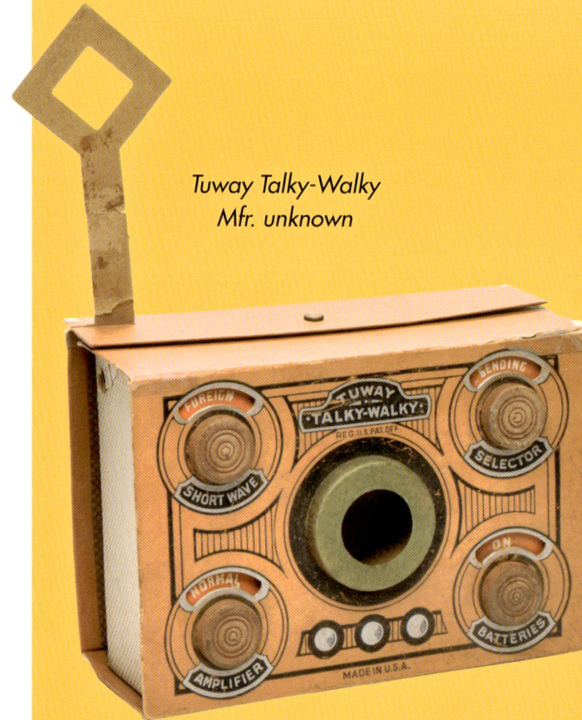

Tuway Talky-Walky
Mfr. unknown

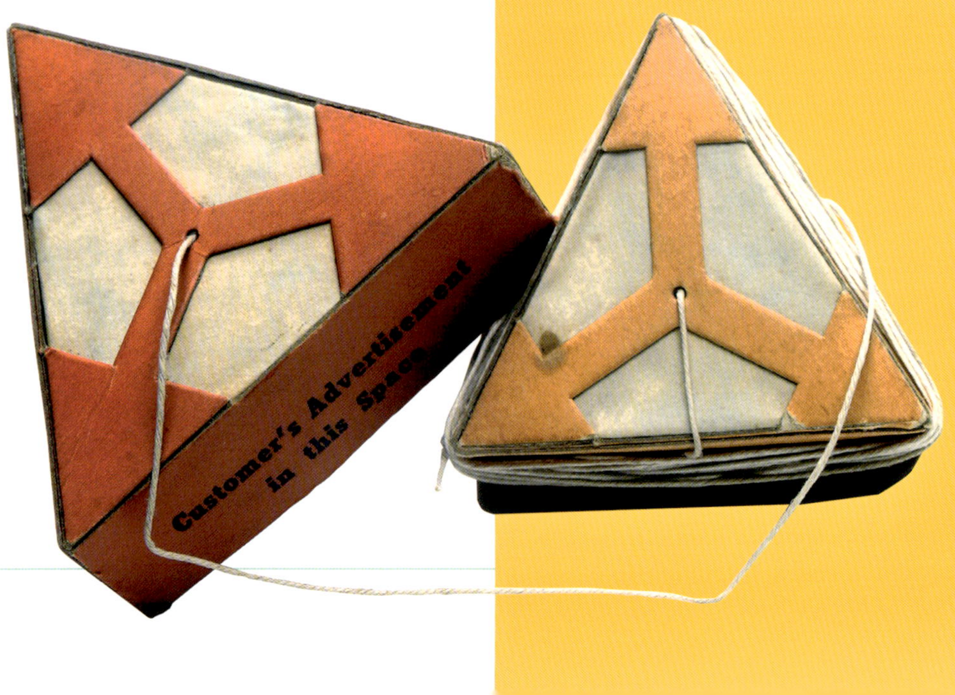

Vibra-Phone advertising premium
Mfr. unknown

Real Talkie Junior Commando Phone
General Wood Work, USA

Turning the crank produced the most satisfying metallic "doink."

Walkie Talkie Radio Phone
Tudor Metal Products, USA

Remco small and large walkie-talkies
Remco Industries, USA
(their first walkie-talkies)

Buck Rogers Super-Sonic Two-Way Trans-Ceiver and box
John F. Dille Company, USA

Buck Rogers Secret Code Book
John F. Dille Company, USA

Gung-Ho Walkie-Talkie Scout
Tudor Metal Products, USA

Assorted WWII lead soldiers
with walkie-talkies

Dick Tracy Crimestopper Club card

Julian D. Newhouse, do you read me?
I have your lost membership card.

Dick Tracy Secret Service
Patrol Code Book

Dick Tracy Detective Wrist Radio
Da-Myco Products, USA

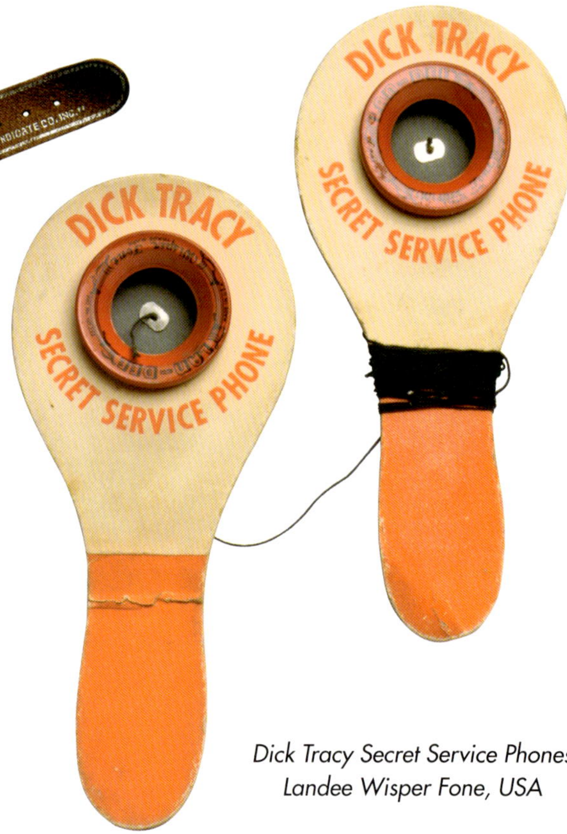

Dick Tracy and his creator, Milton Caniff, get credit for bringing the concept of the miniaturized body-worn communication device to the world. Thank you guys! More versions to come in later years.

Dick Tracy Secret Service Phones
Landee Wisper Fone, USA

Tri-Signal Telegraph Set
Mfr. unknown

Ralston Postal Signal Set
Mfr. unknown

–·–·/–––/––/·// –·/–––/·––//

Communication Set
Modern Toys, Japan

Scene from Plan B From Outer Space *(copy)*
Writer/Director: Ed Wood

1950s

Detail from Jupiter Rocket Launch Pad
Nomura, Japan

Detail from Explorer and Vanguard
Tracking Station
Structoys, Japan

Calling W-1-X-Y-Z
Jimmy Kean and the Radio Spies
Whitman Publications, USA

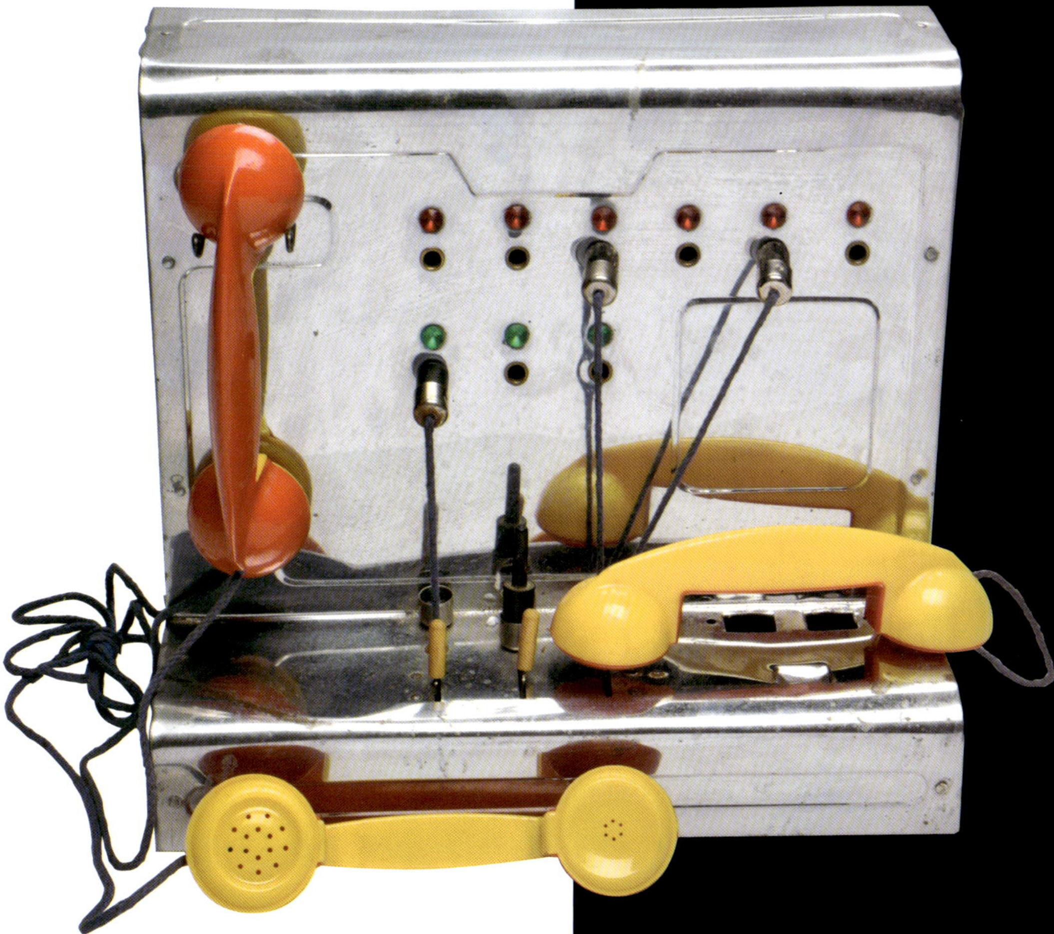

Telephone switchboard
Mfr. unknown

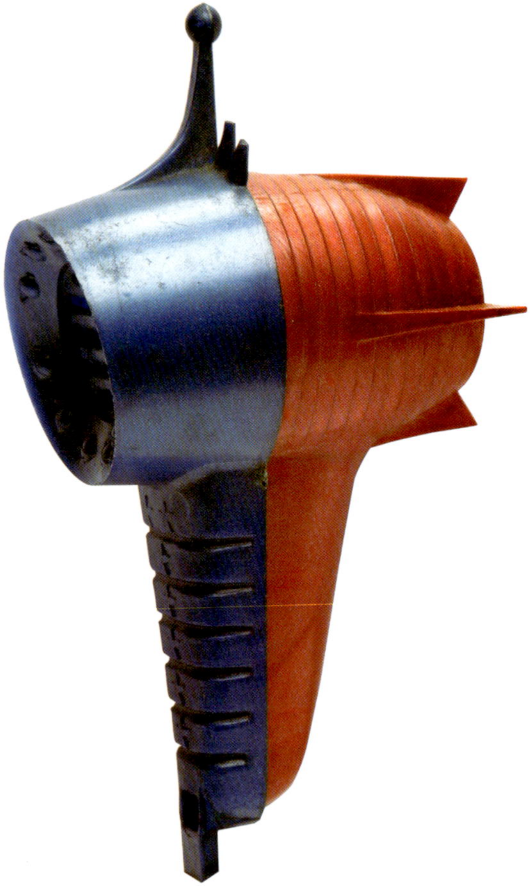

Space Phone Walki-Talki
Selcol, England

Some designs were just too good not to copy.

Space Commander Walkie Talkies
Remco Industries, USA

Tom Corbett 2-Way Phone and box
Vintage Products, Rockwell Productions, USA

Tom Corbett Space Cadet
Inter-Planet 2-Way Phone and box
Designer and Mfr. J. V. Zimmerman Co.
/ Vintage Products, USA

Spacemen
Archer Products, USA

Space Ace Space Phone, box and its Italian-speaking cousin Lone Star Product, England

COSMO PHONE
FUNZIONA SINO A 100 METRI DI DISTANZA
COSMO PHONE
il TELEFONO SPAZIALE
SENZA BATTERIE
SENZA ELETTRICITA'
2 MICROFONI RICE-TRASMITTENTI · AUTO-
GENERATORI DI VIBRAZIONI SONORE ·
CHIAMATA PER MESSA IN ASCOLTO · FUNZIONA
SINO A 100 METRI DI DISTANZA

Ace Brave 3-D pop-up book
Birn Brothers, England

There must be a communication device somewhere in this beautiful pop-up scene.

Publisher unknown

The kind of birthday cards I loved to get when I was a young fellow. Still do, actually.

Card by RustCraft, USA

Dan Dare Walkie-Talkie Set and box
Merit Electronics, England

These folks really understood the importance of great box graphics. And look—push-button planetary selectors! It can't get much easier than that.

VOLUME
WAVELENGTH
TUNING
SECRET
DECODER
TAKE OVER
SWITCH
SATURN
VENUS

TALK HOUSE TO HOUSE • AROUND CORNERS • ACROSS FIELDS
Space Patrol
OFFICIAL WALKIE-TALKIE
SPACE-A-PHONES
PRECISION ENGINEERED
NO BATTERIES
NO PLUG-IN
NO LICENSE
A REAL TWO-WAY SPACE-A-PHONE SYSTEM

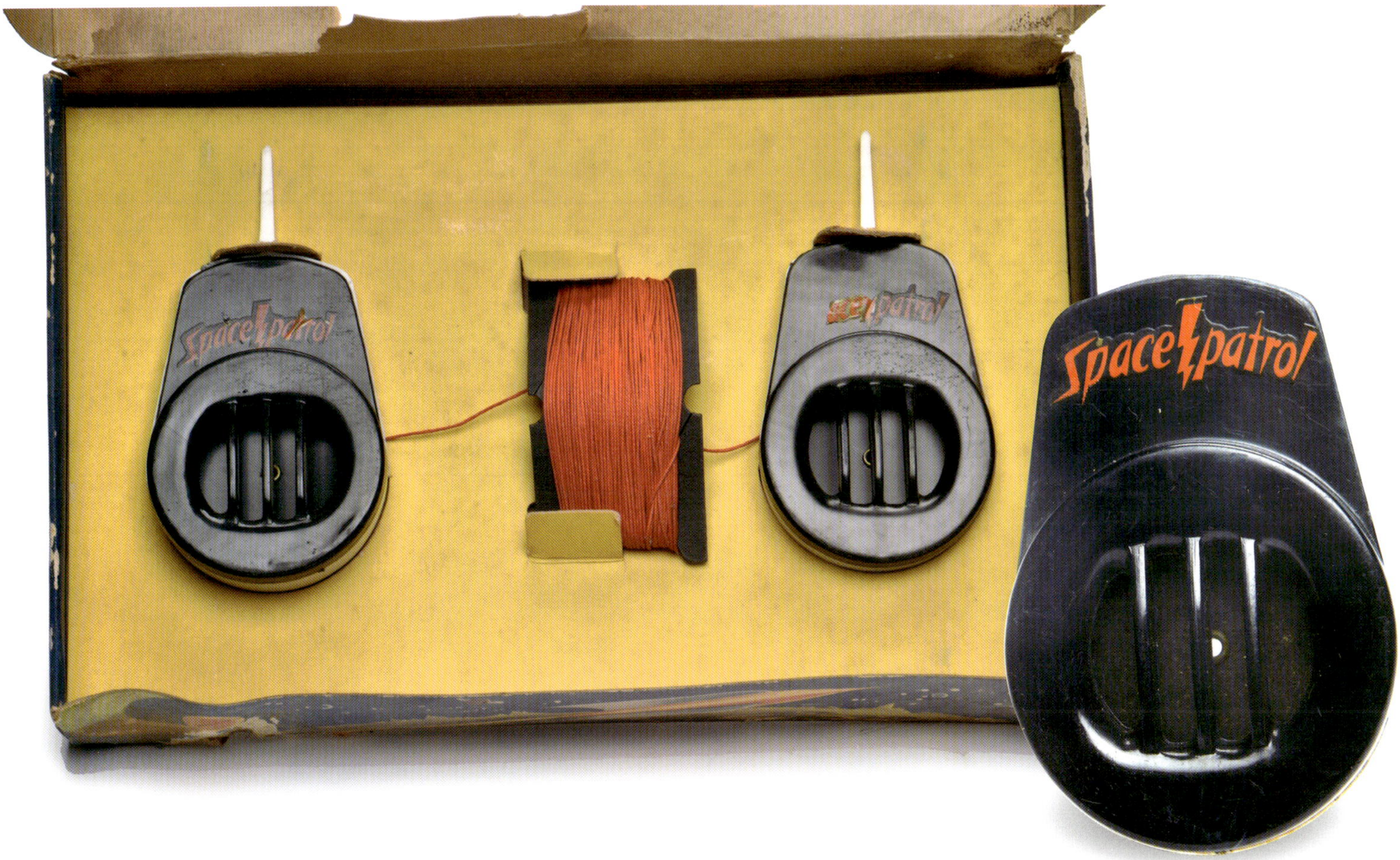

Space Patrol Space-A-Phones and box
Toys of Tomorrow Ltd., USA

Captain Midnight membership book

Space Phones and box
Selcol, England

Zimphone Electronic Walkie Talkie
Futronics Mfg. Co. for Vintage Products Co., USA

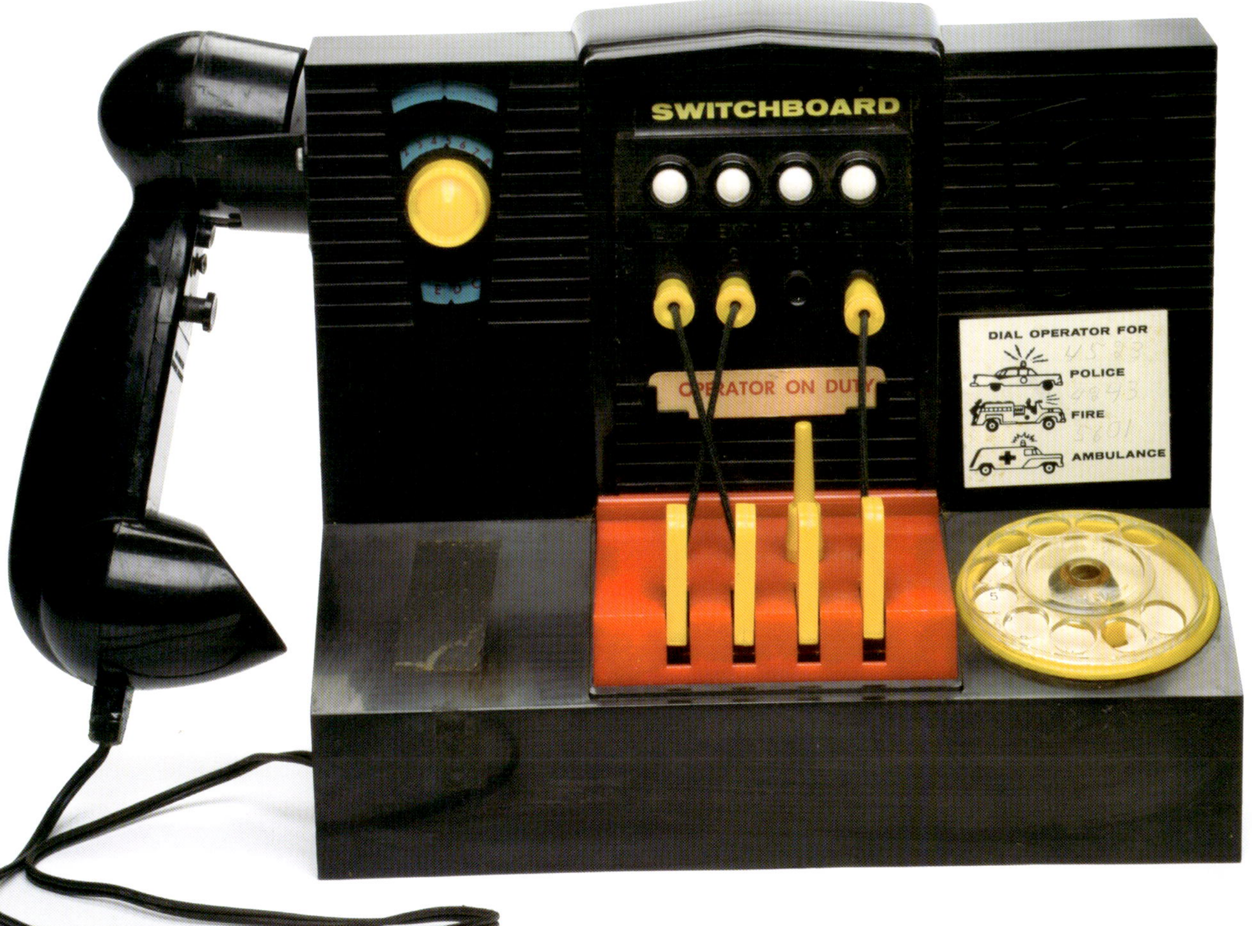

Zimphone operator set
J. V. Zimmerman Corp., USA

Cosmic Phone and packaging variations
U.S. Plastics Co., USA

Captain Video microphone
Mfr. unknown

Electronic walkie-talkie Space Model QX-2
Remco Industries, USA

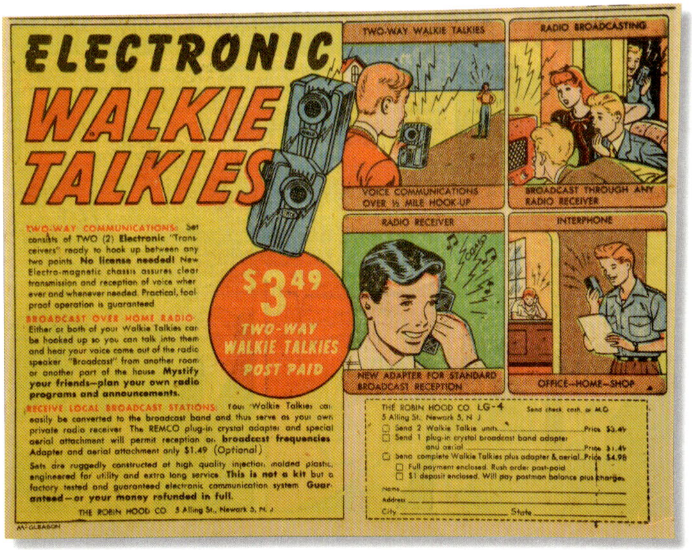

Ad from the Robin Hood Company for walkie-talkie very similar to the Remco version.

Possibly made by Remco as well?

Rin Tin Tin walkie-talkie cereal box ad from the collection of, and photo by, Duane Dimock

Buck Rogers walkie-talkie set with alternate color version. Remco Industries, USA

Electronic Walkie-Talkies
Remco Industries, USA

Yet another packaged version of their classic design.

Raltson Straight Shooters Telegraph Set and Tom Mix Straight Shooter premium also from the Ralston Cereal Company, USA

Who knew Tom Mix hid a secret walkie-talkie in his boot?

Vibra Planet Space Walkie-Talkies Space Phone Set
Vintage Products Co., USA

I love the totally misleading promise. "No electric wires. No batteries." (But did we mention there's 50 feet of string?) Plus it can "broadcast both music and songs." Amazing!

Amateur Broadcaster
Mfr. unknown

Beautiful art deco styling.

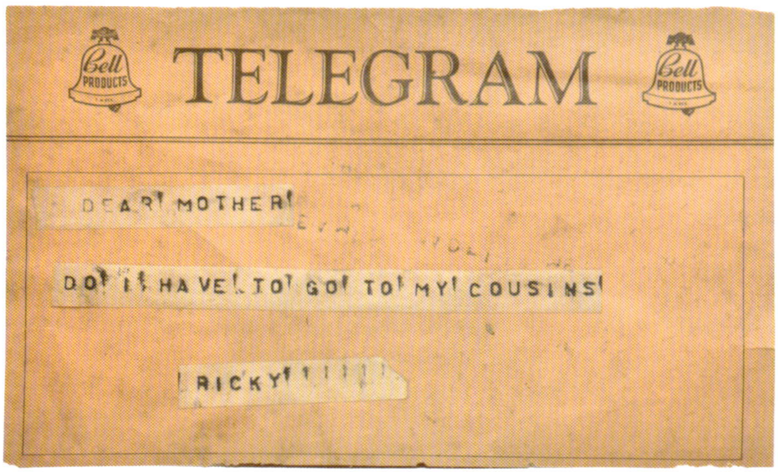

Bell Tele-Typer and telegram
Bell Products, USA

History does not record whether Ricky's transmission was received or his question answered.

Paper signal reflector flasher
Mfr. unknown

Walkie-Talkie premium
Mfr. unknown

I was always thrilled to get these little brown boxes from St. Louis or Battle Creek, Michigan. I still am!

Buck Rogers Interplanetary Space Phone,
assembled from Sylvania Halo Light TV premium kit

The kit contained more than a dozen other wonderful Buck Rogers paper items.

Detail from Buck Rogers original comic art.
Artist: Rick Yeager

Unidentified rocket cockpit photo

Isn't it a great one?

Wrist Communicator
Mfr. unknown

1960s

Zimphone handset
J. V. Zimmerman
Brumberger Model 250
Brumberger Co., USA

This handset seems to be the same one used on the Zimphone Operator Set manufactured by J. V. Zimmerman. Not sure how to explain the double branding.

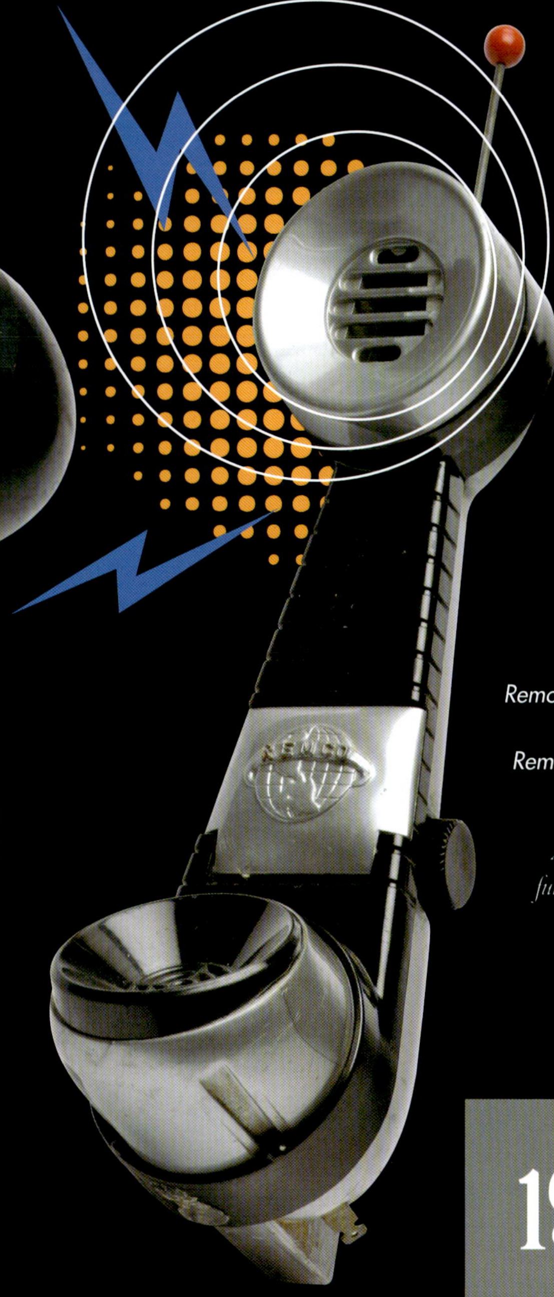

Remco Magnet Powered
Walkie Talkie
Remco Industries, USA

Another beautifully futuristic Remco design

Remco Industries ad

Remco Electronic Mobile
Loud Speaker
Remco Industries, USA

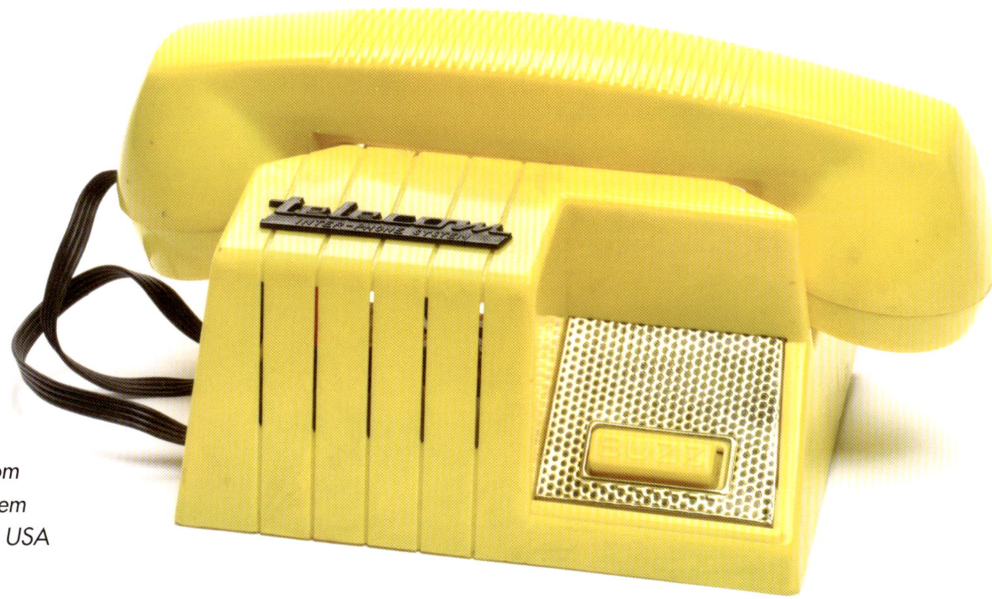

Remco Tele-Com
Interphone system
Remco Industries, USA

Crystal Radio Set
Radiocraft, USA

Space Phones
Mfr. unknown, Hong Kong

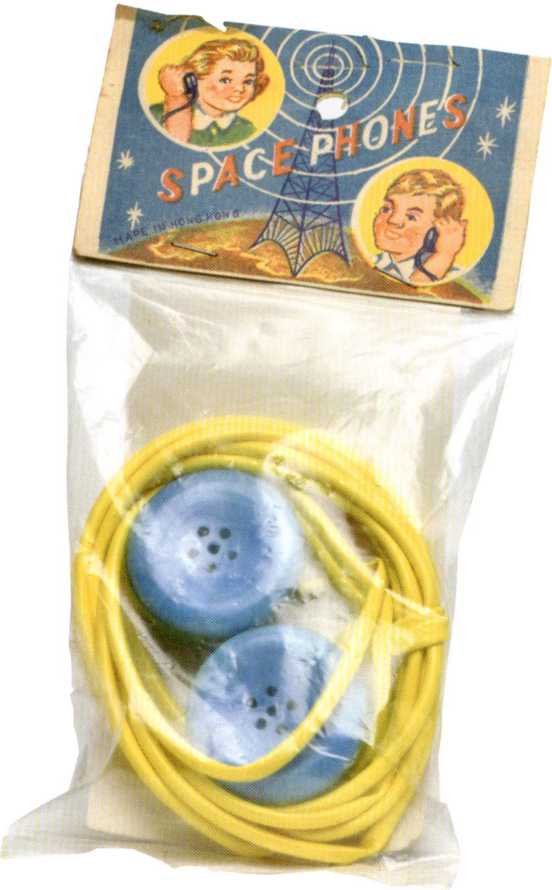

Kazoo Space Phone
Mfr. unknown, Hong Kong

Original Batman coloring-book illustration
Jason Studios, artist is possibly Dan Spiegle

Boy Scouts of America SOS Signaler and box
National Supply Service Division, USA
Collection of Melinda Woods

Radarphon and box
Geobra, West Germany

What the box lacks in cool, the Radarphone more than makes up for in brilliant futuristic style. Simply gorgeous.

Radarscope
Product Miniature Co., USA

Space Station Morse Code Signalling Set
Mfr. W, Hong Kong

Commando Walkie Talkie
Gabriel Industries, USA

Archer Space Patrol Walkie Talkie
Archer Manufacturing
for Radio Shack, USA

Transistor Walkie Talkie
Heathkit, USA

Remco Combat Field Walkie Talkie
Remco Industries, USA

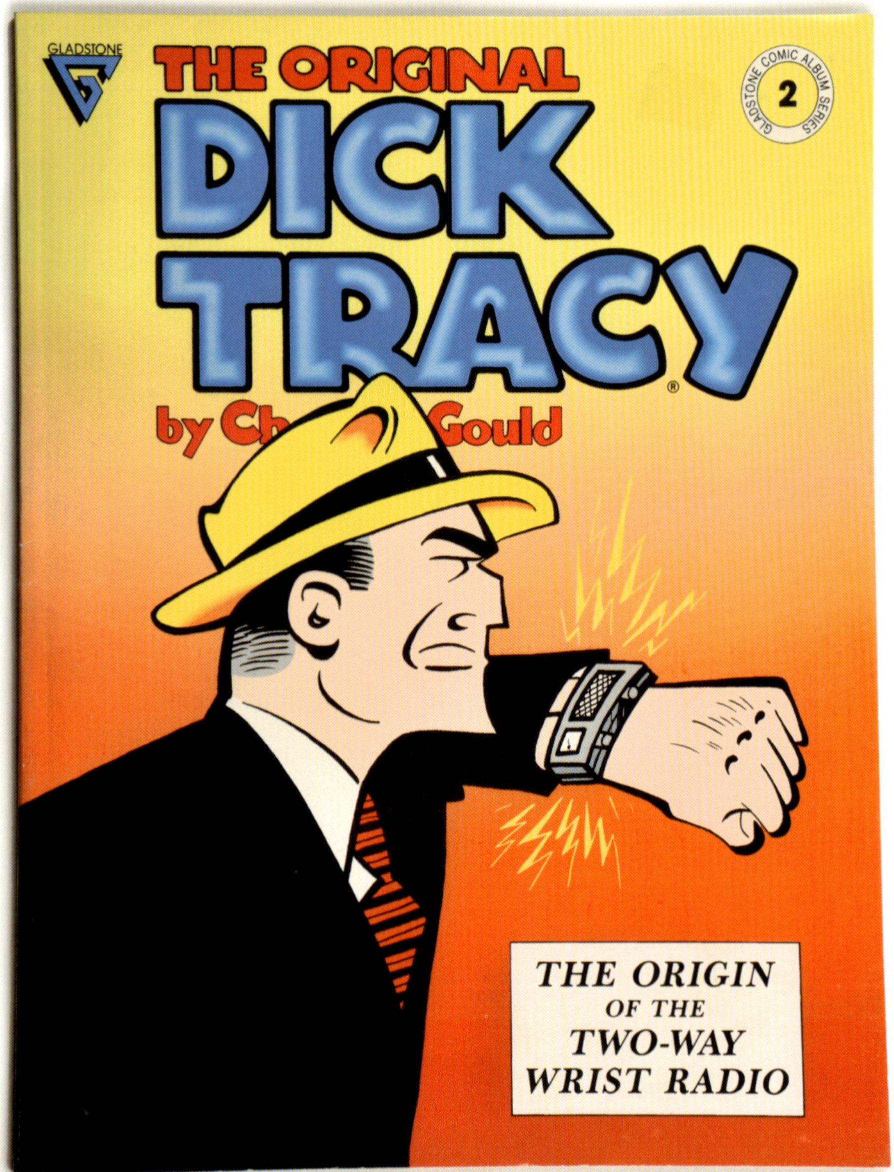

The Original Dick Tracy:
The Origin of the Two-Way Wrist Radio
Gladstone Publishing

Dick Tracy Wrist Radios
Remco Industries, USA

Dick Tracy TV Detective Club card

Dick Tracy original illustration panel, 1971
Artist: Chester Gould

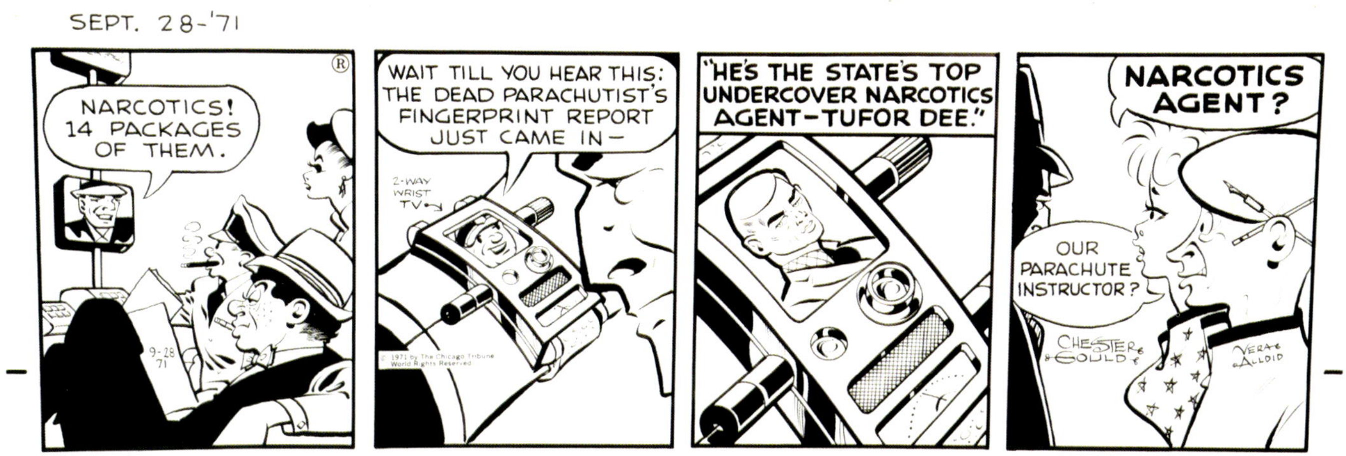

Warehouse 13 Farnsworth
Handmade by Steve Joy,
Santa Barbara, California

A beautifully created piece with voice sound chip.

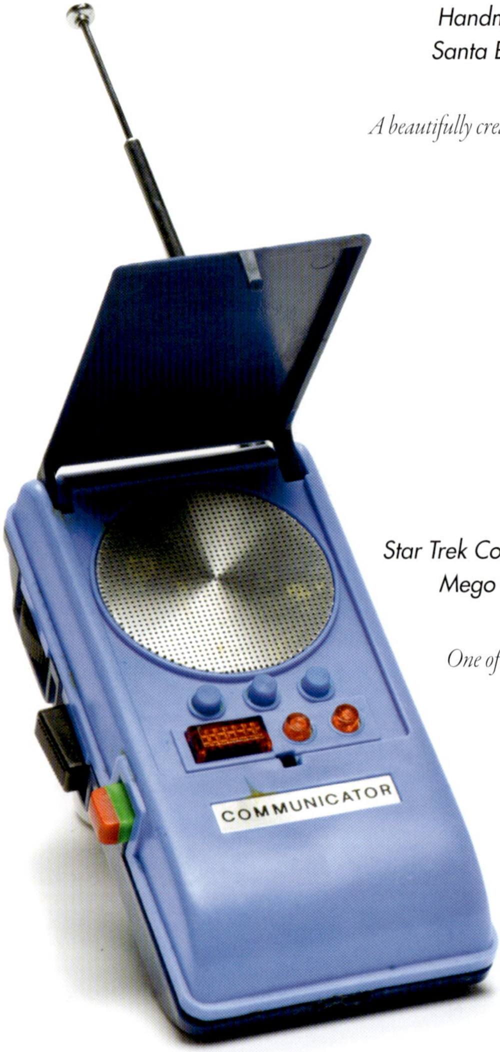

Star Trek Communicator, circa 1970
Mego Corporation, USA

One of many, many versions.

Space Radio
Mfr. unknown, Hong Kong

Infrared Astro Phone
Infrared Industries, USA
From the collection of, and photo by, Eric Wrobbel

A giant leap forward from talking through a string, no?

Astro Witch original art
Artist: Chad Maupin, Big-Bot Design,
Fayetteville, Arkansas

Check out her antennae!

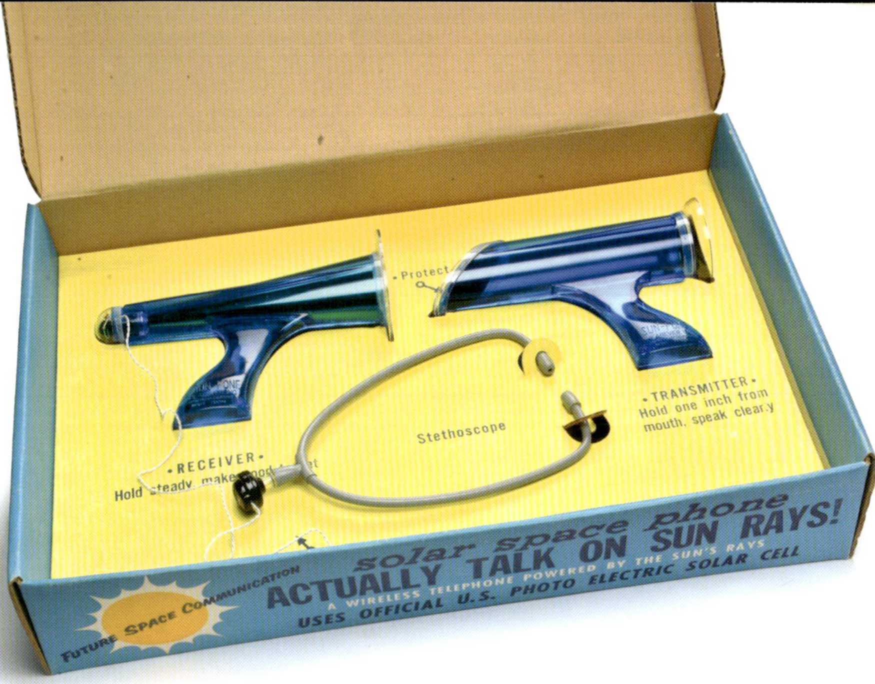

Solar Space Phone
Hear-Ever Co., USA

Beautiful design and nice technology breakthrough.

Caravelle Transmitter Receiver
Remco Industries, USA

Yet another fabulous Remco design.

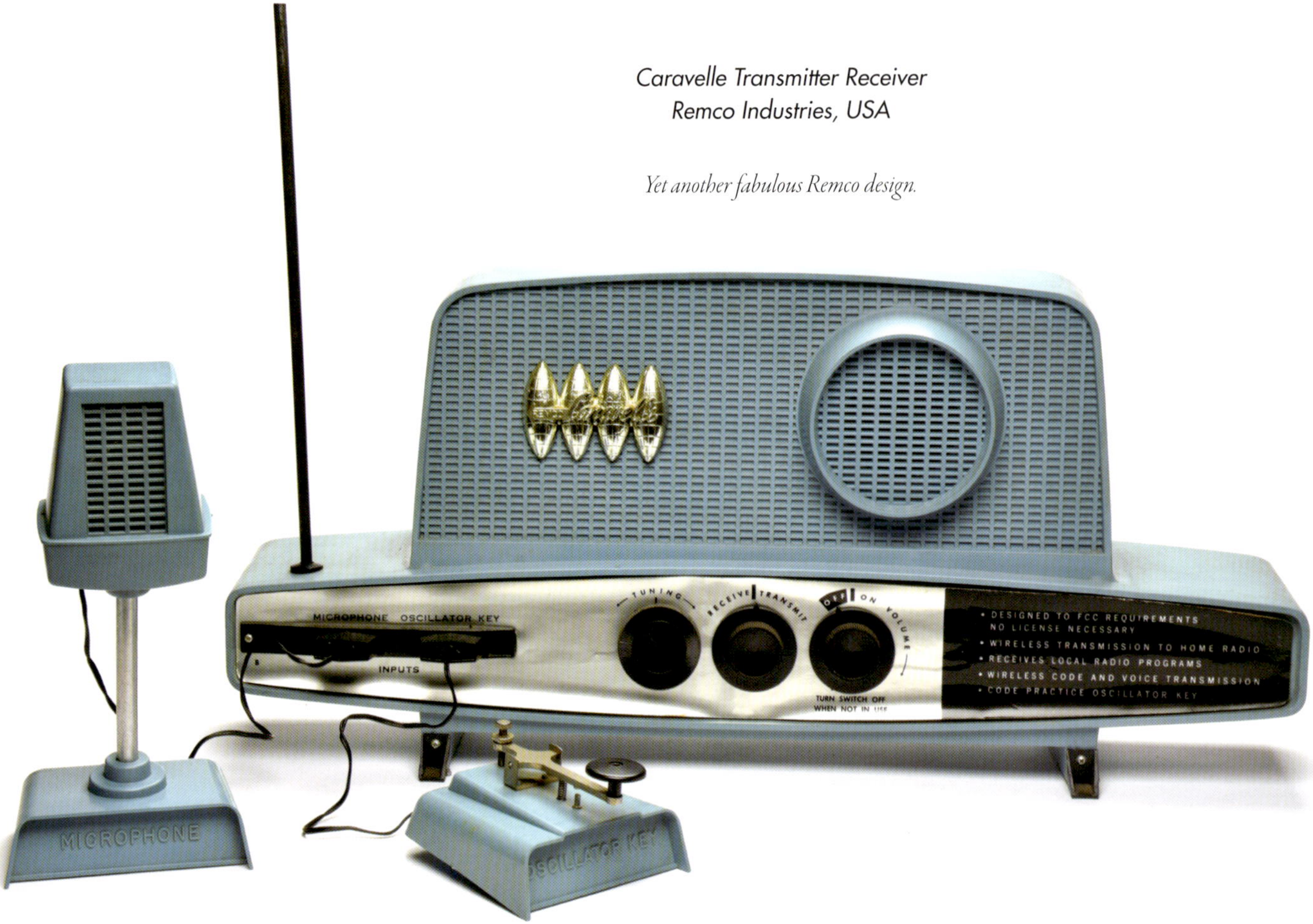

Assorted space and military communication
guys and equipment
Mfrs. unknown

Illustration from Space Ship to the Moon
Rand McNally Book-Elf Books, 1952
Artist: A. K. Bilder

This kid's illustration has everything. Futurism, steampunk, and whimsy. Perfect!

Apollo Command Module microphone
General Electric Co., USA

This was not actually meant as a toy,
but it sure made a great one.

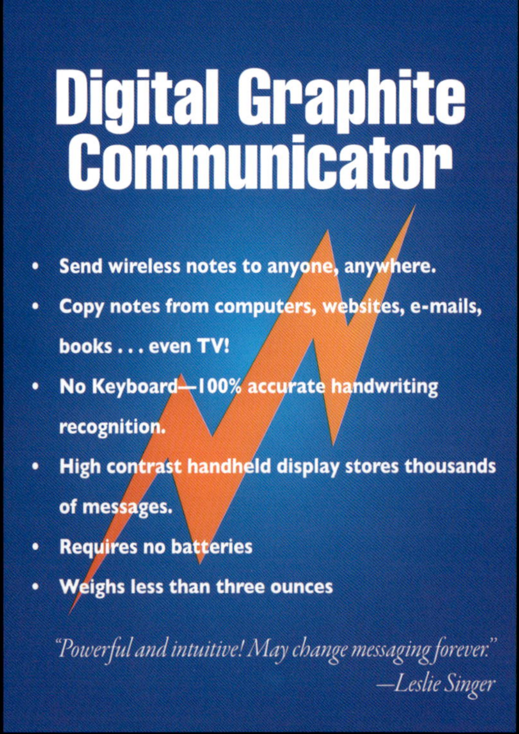

Singer Digital Graphite Communicator and box.
Artwork: Patti Woods Rye; Copy: Leslie Singer

When Palm Pilots were the newest rage in the '90s, I believe I outdid them.

Instructions:

1. Grasp stylus between digits with graphite interface down.
2. Transcribe any alpha-numeric symbols—even equations, graphs, and drawings—by traversing graphite interface across work pad surface.
3. Rotate vector of stylus to access delete feature.
4. Extract pad surface to "send" message. A new pad surface appears automatically. (Please recycle previous message surface when possible.)

Caution: Do not immerse in water or operate near open flame.

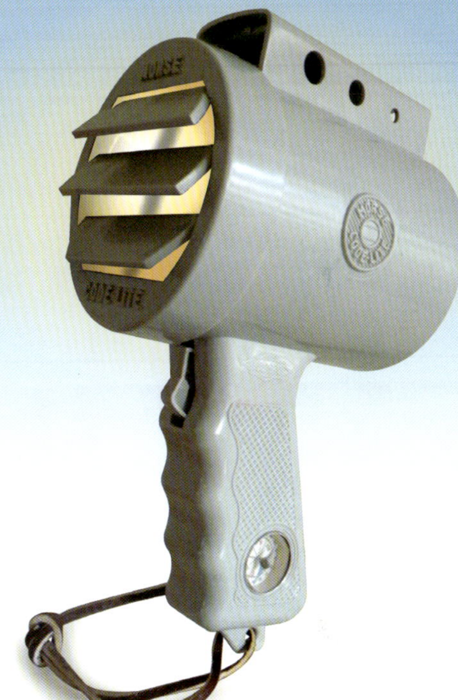

Navy Blinker Code Lite
Hasbro (Hassenfeld Bros.), USA
Collection of Melinda Woods

Children's birthday card

Children's valentine card

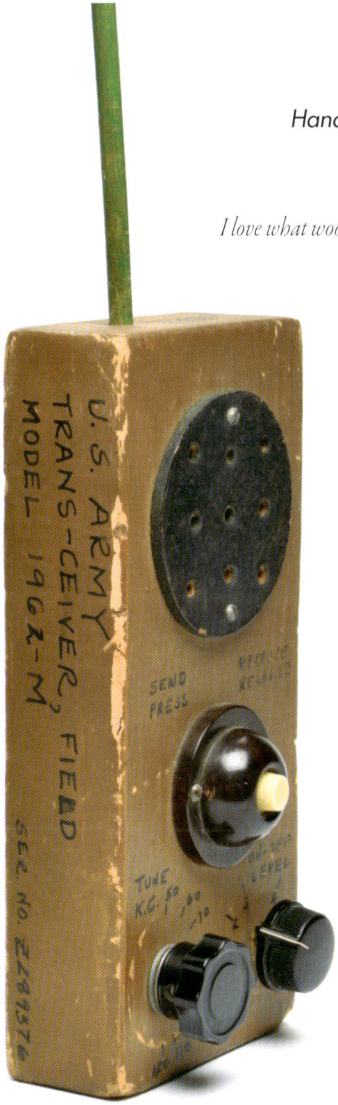

Handmade walkie-talkie toy
Maker unknown

I love what wood and imagination have created here.

Sask-A-Phone
Mfr. unknown

Handmade note pad communicator
Maker unknown

Another creative communication device.

Armellini Radio Company Strato-Spheric Trans-Ceiver

A beautiful one-of-a kind set and box made for me by my fellow communicator, Rick Armellini.

More than we ever imagined.
What's next?

eighty-five

Ranger Rays Rocketeers
Stella Roberts Fonts

HEY KIDS! TURN THE PAGE TO BUILD YOUR OWN SINGER COMMUNICATOR.

NO WIRES.
NO BATTERIES.
NO INTERPLANETARY OR IONIC INTERFERENCE.

SINGER COMMUNICATIONS WALKIE - TALKIE

MODEL # XIB / PURE 300

DESIGN AND PAPER ENGINEERING BY RICK ARMELLINI

1 - COPY ON TO CARD STOCK.

2 - CUT OUT THE LARGE WALKIE-TALKIE AT RIGHT IN ONE PIECE.

3 - USING THE SMALL SAMPLE BELOW AS A GUIDE, CUT ORANGE SLOTS D, E, F, G.

4 - CREASE & FOLD DOWN ALONG GREEN DASHED LINES.

5 - BEND TABS D & E TOWARD THE CENTER, LOCKING SLOTS D & E TOGETHER.

7 - BEND TAB AA UP & OVER TABS D & E.

8 - BEND TAB A DOWN & USE DOUBLE-SIDED TAPE TO STICK IT TO TAB AA.

9 - REPEAT FOR TABS F, G, C & CC.

10 - FOLD TAB B & USE DOUBLE-SIDED TAPE TO STICK IT TO TAB E-G.

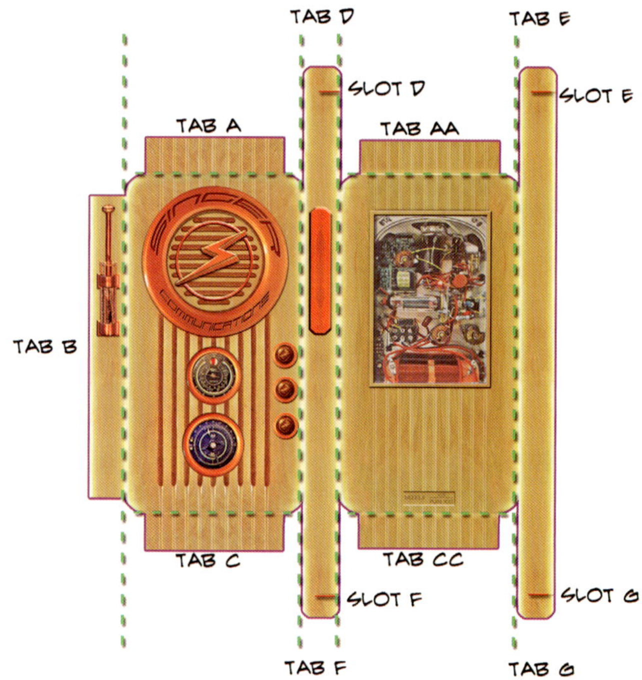

SINGER
COMMUNICATIONS
MODEL #
X1B
PURE 300

Price guidelines

The prices listed in this book are according to the author's experience and are for information purposes only. Prices will vary with demand, condition, locale, knowledge, desire of the buyer and seller, and other factors. Buying and selling are at the reader's risk, and neither the author nor publisher shall be held responsible for losses that may occur through use of this book or selling of any items, or due to typographic or other errors.

Page	Item	Good–Mint *Range: Good single piece to mint with packaging as of publishing date.*
4	Space Phone	\$10–\$65
5	Electronic Walkie Talkie / Mfg. unknown	\$25–\$95
8	Tom Corbett puzzle / Saalfeld	\$8–\$25
10	Walki-Talki DCMT Ltd./ Lone Star Products, England	\$25–\$150
11	Walkie Talkie Phone	\$125
13	Remco Radio Station	\$35–\$100
15	Biekens Biscuits card	\$3–\$5
16	Radio-Phones	\$10–\$25
17	Biekens Biscuits card	\$3–\$5
18	Two-Way Radio	\$10–\$60
18	Radio Tinker	\$15–\$100
19	Little Wonder Mic	\$15–\$50
20	Flash Gordon 2 Way	\$20–\$150
21	Phon-o-Mike	\$10–\$25
21	Pillsbury's Tel-A-Phone	\$10–\$30
21	Tums walkie talkie	\$10–\$25
21	Micro-Fone	\$10–\$30
21	Radio-Mike	\$10–\$25
21	Coco-Wheat Mike-O-Phone	\$10–\$25
22	Telephones	\$10–\$25
22	Vibra-Phones	\$10–\$25
22	Tuway Talky-Walky	\$50–\$90
23	Remco Walkie-Talkies	\$25–\$50
23	Real Talkie Junior Commando	\$35–\$60
23	Walkie Talkie Radio Phone	\$35–\$50

Page	Item	Good–Mint *Range: Good single piece to mint with packaging as of publishing date.*
24	Buck Rogers Super-Sonic Two-Way Trans-Ceiver	$50–$500
24	Buck Rogers Secret Code Book	$50–$125
25	WWII Soldiers, ea.	$10–$35
25	Gung-Ho Walkie-Talkie Scout	$15–$35
26	Dick Tracy Secret Service Patrol Code Book	$15–$30
26	Dick Tracy Crimestopper Club card	$5–$25
26	Dick Tracy Detective Wrist Radio	$50–$225
26	Dick Tracy Secret Service Phone	$20–$65
27	Tri-Signal Telegraph Set	$10–$50
27	Postal Telegraph Signal Set	$10–$50
28	Communication Set	$50–$350
30	Calling W-1-X-Y-Z	$5–$25
31	Telephone switchboard	$15–$45
32	Space Phone Walki-Talki	$15–$75
33	Remco Space Commander	$15–$75
34	Tom Corbett Space Cadet Inter-Planet 2-Way Phone	$35–$150
35	Archer Spacemen, ea.	$5–$10
35	Tom Corbett 2-Way Phone	$50–$150
36	Space Ace Space Phone	$50–$150
37	Cosmo Phone	$50 – $150
38	Ace Brave pop-up book	$35–$85
39	Birthday cards, ea.	$5–$8
40	Dan Dare Walkie-Talkie set boxed	$100 –$250
42	Space Patrol Space-A-Phones, boxed	$100–$175
44	Captain Midnight membership book	$20–$35
45	Space Phones, boxed	$75–$100
46	Zimphone Electronic Walkie Talkie, boxed	$75–$125
47	Zimphone operator set	$25–$100

Page	Item	Good – Mint *Range: Good single piece to mint with packaging as of publishing date.*
48	Cosmic Phones, on card	$35–$50
48	Cosmic Phones, boxed	$50–$75
49	Captain Video microphone	$65–$175
50	Electronic QX-2 Walkie-Talkies	$15–$100
51	Electronic Walkie-Talkie ad	$5–$10
51	Rin Tin Tin Walkie-Talkie, Shredded Wheat box only	$50–$100
52	Buck Rogers Walkie Talkies and box	$100–$135
53	Remco Electronic Walkie Talkies, ea.	$30–$45
54	Tom Mix Straight Shooter string phones	$15–$25
55	Vibra Planet Space Walkie-Talkie ad	$5–$10
55	Strato-Scout Walkie-Talkie Space Set, ea.	$15–$25
55	Strato-Scout Walkie-Talkie Space Set in package	$85–$150
56	Amateur Broadcaster	$50–$175
57	Bell Tele-Typer	$75–$150
58	Walkie-Talkie premium	$15–$40
58	Paper signal reflector flasher	$5–$10
59	Buck Rogers Interplanetary Space Phone, ea.	$15–$25
59	Buck Rogers original illustration (Yeager)	$125–$250
60	Wrist Communicator	$5–$10
61	Zimphone handset	$10–$20
61	Remco Magnetic Powered Walkie Talkie	$15–$25
62	Remco ad	$5 – $10
63	Remco Electronic Mobile Loud Speaker	$40–$100
63	Remco Tele-Com Interphone	$20–$90
64	Crystal Radio Set	$25–$100
65	Kazoo Space Phone	$2–$8
66	Batman original art (Spiegle?)	$35–$60
67	Boy Scouts of America SOS Signaler	$20–$50

Page	Item	Good–Mint *Range: Good single piece to mint with packaging as of publishing date.*
68	Geobra Radarphon	$25–$250
69	Radarscope	$45–$150
70	Space Station Morse Code Signalling Set	$20–$55
71	Commando Walkie Talkie	$25–$45
71	Archer Space Patrol Walkie Talkie	$10–$20
71	Heathkit Transistor Walkie Talkie	$25–$50
71	Remco Combat Field Walkie Talkie	$20–$40
72	*The Original Dick Tracy*	$5–$10
72	Dick Tracy Wrist Radios	$25–$50
73	Dick Tracy TV Detective Club card	$15 – $40
73	Dick Tracy illustration (Gould)	$150–$300
74	Star Trek Communicator	$15 – $25
74	Warehouse 13 Farnsworth (Joy)	$75 – $125
74	Space Radio	$10 – $15
75	Infrared Astro Phones	$25 – $95
76	Astro Witch original art (Maupin)	$75 – $100
77	Solar Space Phone	$35 – $95
78	Remco Caravelle Transmitter Receiver	$65 – $100
78	Assorted space and soldier figures	$3 – $25
79	Birthday card	$5 – $8
79	Apollo microphone	$20 – $45
80	Digital Graphite Communicator (Singer)	N/A
81	Navy Blinker Code Lite	$20 –$65
82	Greeting cards, ea.	$5 – $10
83	Handmade walkie-talkie	O.O.A.K.*
83	Handmade note pad	$8 – $15
83	Sask-A-Phone	$3 – $15
84	Strato-Spheric Trans-Ceiver (Armellini)	O.O.A.K.*

* One of A Kind

Signaling my gratitude to these fine sources

Steve Joy's great stuff can be found on eBay and Etsy under Stickbass and at his website:
www.joy-stick-bass.com

Chad Maupin is a wonderful artist and the master of Big-Bot Design. Check out his very cool stuff at:
www.bigbot.storenvy.com

Eric Wrobbel is a collector of many wonderful things. He has produced several books on old toy walkie-talkies, crystal radios, and other vintage-technology collectibles, which you can find at www.ericwrobbel.com—a visual treat featuring dozens of collections and hundreds of photos.

And to Pete Schiffer, Jesse Marth, and everyone at Schiffer Publishing for all their assistance, contributions, and allowing me to send this fun project out into the universe.

Leslie

Leslie Singer is an advertising creative director who self-medicates by collecting vintage space toys, original sci-fi illustrations, and other retrofuture graphics.
He lives in Little Rock, Arkansas.

Willie Allen is a commercial and former United Press International photographer whose work includes many of the iconic photos from the Kennedy assassination. He is a longtime personal pal and all-around great human.

Willie

Patti Woods Rye is an Arkansas-based art director who brings her creative vision to advertising projects and to the Pig Trail Trading Company, her own line of whimsical clothing products.

Patti